George Summer

Recollections of service in Battery

First Rhode Island Light Artillery

George Summer

Recollections of service in Battery
First Rhode Island Light Artillery

ISBN/EAN: 9783337271008

Printed in Europe, USA, Canada, Australia, Japan

Cover: Foto ©berggeist007 / pixelio.de

More available books at **www.hansebooks.com**

SOLDIERS AND SAILORS HISTORICAL SOCIETY

—OF—

RHODE ISLAND.

PERSONAL NARRATIVES.

FOURTH SERIES,

Nos. 11 to 20.

1891-93.

PERSONAL NARRATIVES

OF EVENTS IN THE

WAR OF THE REBELLION,

BEING PAPERS READ BEFORE THE

RHODE ISLAND SOLDIERS AND SAILORS

HISTORICAL SOCIETY.

FOURTH SERIES – No. 11.

PROVIDENCE:
PUBLISHED BY THE SOCIETY.
1891.

𝕿𝖍𝖊 𝕻𝖗𝖔𝖛𝖎𝖉𝖊𝖓𝖈𝖊 𝕻𝖗𝖊𝖘𝖘:

SNOW & FARNHAM, PRINTERS,

37 Custom House Street.

1891.

RECOLLECTIONS OF SERVICE

IN

BATTERY D,

FIRST RHODE ISLAND LIGHT ARTILLERY.

BY

GEORGE C. SUMNER,

[Late of Battery D, First Rhode Island Light Artillery.]

PROVIDENCE:
PUBLISHED BY THE SOCIETY.
1891.

RECOLLECTIONS OF SERVICE

IN

BATTERY D,

FIRST RHODE ISLAND LIGHT ARTILLERY

———◆———

THE spring and summer of 1861 was full of excitement for the young men of that day First, rumor of war, then actual war stirred their patriotism to its very depths. Then the enlistments began, and soon every armory was filled with men and boys full of excitement; soldiers paraded day and night; the beating of drums could be heard at all times, and one could hardly pass along the streets without meeting companies out for drill, while train after train passed through the city filled with soldiers off to the war.

I was a boy of seventeen in those days, with per-

haps the average amount of patriotism. At all
events I imbibed my full share of the excitement, and
it was with considerable effort that I repressed it suf-
ficiently to prevent my enlisting, which I was unsuccessful in doing until the 2d of September. On
that day a friend and myself strolled over to the
Marine Artillery armory, on Benefit Street, and while
there it was announced that Battery D needed only
about a dozen men to finish its complement. John
said to George, " What do you say? " George re-
plied, " It is a go ; " and down went our names for
three years or the war.

The writer went immediately home and informed
his people of his determination to go to the war, si-
lencing all opposition by announcing that the deed
had been done, as he had placed his name on the roll.

We were mustered in on September 4th, and went
to Camp Ames near Pawtuxet, where we were drilled
in marching for a day or two. On the 13th boarded
the cars, and were taken to Stonington, leaving that
night on the boat for Elizabeth City, N. J., where
we took the cars direct for Washington, arriving on
the 15th, and went directly to " Camp Sprague,"

where we remained until October 12th. During this time we drew our guns and horses, did lots of drilling, had several reviews,—one by General Scott, —and numerous opportunities to look about the city. I used frequently to go to the Capitol, climb to the top of the unfinished dome, and take a look over into Virginia, hoping to catch a glimpse of the rebels, but as my vision was limited to five or six miles, instead of the necessary fifty, I was of course unsuccessful.

On October 12th we were ordered to pack up and moved through Washington over Long Bridge into Virginia, marching five or six miles, and went into camp at a place called Hall's Hill. On the 14th we drew A tents, and pitched camp in the woods on a side hill.

The only thing I remember particularly about this camp is the immense camp-fires we used to have. The nights were rather cool, and we used to build these fires and then fell the trees over on to them, and as fast as the limbs burned off would pile them on, and soon had a fire that half the members of the battery could get around, and roast one side while the other froze.

On the 17th we moved our camp some three miles
to Upton's Hill, and on November 2d, to Munson's Hill. We remained here for a week or ten
days, and then moved into what was thought to be
a better location, and began to build our winter's
camp. We parked the battery in regular style, pieces
in front, caissons in the rear, and on either side of
these we built stables for our horses, by first building
a framework of poles and covering the top and sides
with pine boughs. Our tents were pitched on a line
with the stables, extending nearly to the officers'
quarters, which left quite a commodious parade
ground between the latter and the battery-park for
inspections, guard-mountings, etc. These were Sib-
ley tents, circular in form and quite large, with a
stove in the centre. I have forgotten whether we had
one or two to a detachment ; at any rate we had plenty
of room. The bunks were built large enough to ac-
commodate two, and were filled with straw, and as
each man had two blankets, by bunking together we
could lay two under and use two over us, with our feet
towards the fire. The man from the tent who happened to be on guard made it his duty to see that

the fire did not go out on cold nights. You may be sure that we slept just as comfortably as had we been on a feather bed at home.

We had as neighbors in this camp Battery B, Fourth United States Artillery, Captain John Gibbon, First New Hampshire, Captain Gerrish, and a Pennsylvania battery, Captain Durrell. Captain Gibbon of the regular battery, had command of the post. Quite a rivalry existed between the regular and volunteer company, and I recollect that some tall hustling used to be done to prevent their beating us in moving out of park, after "Boots and Saddles" had been sounded for drill or inspection. We seldom got left, but frequently had narrow escapes, and well do I remember how anxious we used to be on the left piece, sometimes, until we heard the order, "Right piece, forward!" fearing that it might please the fancy of the captain to move the battery out of park, left in front; as we were not quite ready, some kind-hearted cannoneer, at that moment, was finishing what the driver had not had time to do before the order to mount.

We used our time that fall and winter in drill, inspections, learning to ride, and the manual of the piece. I don't think I had ever mounted a horse before I became connected with Battery D, and well do I remember my experience in mastering the art of horsemanship. The method used for our instruction was heroic, with the single exception that we were allowed to put a blanket over the backbone of the horse, no saddles, no bridles were allowed. I should be sorry to attribute a wrong motive for the blanket consideration, but I am obliged to say that it was my opinion then, and is still, that the privilege was allowed us more from fear that without this protection we were liable to be incapacitated for guard and other duties, which we were obliged to do on foot, rather than our personal comfort.

We were obliged to control the horse as best we could with the halter, which was practically no control at all. The horse that I took my first lesson on was not a good saddle horse. He had only one easy gait, that was a walk. His trot was fearful. He could easily lift me four or five inches from his back every time he put his feet down ; his running was not much better ; lope he could not, and so sometimes when the rest of the battery were loping along easily, I was being pounded almost to death by his constantly changing gait from trot to run or run to trot, as he fell behind or gained on the rest of the column.

Many of us had the bad habit of holding on with our heels. We were cautioned time after time that we must not do so, but use our knees for that purpose, but the very next time we would forget and commit the same error ; so some fertile brain among the officers conceived the horrible plan of placing spurs upon the heels of those who were troubled with this forgetfulness. I made up my mind that I would be sure and remember about those spurs, and succeeded very well so long as we remained at a walk ; but when ordered to trot, and I commenced bounding all over that horse's back, I forgot, and hugged my horse with my heels.   The effect was electrical.   My horse darted ahead, and hitting the one in front, whose heels I was just able to dodge, started down the off side of the line on the dead run.   I did some fine dodging on that trip. Seventy-five per cent. of the horses in that line tried to hit me, but not one succeeded. All this time my poor horse was asking me, I suppose, to let up with those spurs; but I was so confused and astonished that I did not catch his idea, and he, despairing of my taking a reasonable view of the thing, and having reached a fence which barred his further progress, invited me in such an intelligent manner to dismount that I understood him, and complied, not in the graceful manner that I should have done if my will had governed it, but over his head and the fence into an adjoining field, all in a heap, and it was not until I had rubbed the pain out of my bruised head, arms and body, that I fully realized that the cause of my trouble had been those spurs.

Occasionally we would go out for target practice, and I remember on one occasion we had been firing in a direction, for example, towards the south, when soon up rides a colonel and desires to know why we were shelling his camp. The captain informed him that we had not been shelling in his direction at all, and pointed in the direction that we had been firing. The colonel said he did not care if we had been firing south, our shells had been going west. That lot of ammunition was speedily condemned.

About this time General McDowell, who commanded our corps, began to have his splendid reviews and sham fights. I had an experience at one of these sham fights that I must relate. The general would have the whole corps out, running them all over those plains, fighting an imaginary enemy, and firing blank cartridges. On one of these occasions I was driving the wheel team caisson, and we had been having a hard fight, when the general suddenly discovered the enemy had got around on our flank, and he gave orders to change our front, thereby giving the battery a run of about a mile, which would have been pleasant enough to us if it had not been for the fact that a large part of the way lay through what had been woods when the corps first went into camp, but it had been cut off by the soldiers, and the stumps left by them were of the usual irregular height that a soldier always left. No true soldier would ever bend his back in cutting down a tree, consequently the height of the stump varied according to the height of the man cutting it down. Well, we started on our run, and it was a fearful one for cannoneers. We would strike a

2

stump and they would leave the box heavenward,
meeting it again by the force of gravitation, and they
had all they could do to remain on the boxes. We
made the journey, and arrived at a place where I
thought we could come about, and held in my team,
but my lead driver thought differently, seeing which,
I let my team out again, and I had hardly done so
when around he went, and the swing driver, who
was a green one, instead of swinging out, as he
should have done, did just the other thing, which
brought his traces taut across the pole, and pulled
my team around in spite of all they could do, and to
add to our trouble, there happened to be a large
stump in just the right place for the off wheels to
hit it, which they did with great force, and over went
the caisson, bottom side up, throwing both my horses
and myself badly mixed up with them.

You may imagine my feelings as I saw that caisson going over, knowing that in the chests there were
a dozen or more shells of the Parrott pattern, which
were exploded by a percussion cap affixed to a
plunger inside the shell, and needed only a severe
concussion to explode them.

The mind works quickly at such times. I expected

they would explode, and that being so near them I
should probably be killed. I thought of home,
friends, and a thousand other things during those
few seconds; but fortunately they did not explode,
although the cases were completely smashed. I
managed to extricate myself from the debris, and
was trying to arrange things, when General McDowell rode up with his grand staff and ordered a company of infantry to assist us in straightening out; but
their help was in reality a hindrance, as they knew
nothing about our work, and the first thing I knew
they had unbuckled every buckle in the harnesses,
and it took longer to buckle up again than to have
straightened it out half a dozen times if left to ourselves; but finally we got fixed and went back into
camp. Captain Monroe told me that night that the
government would probably expect me to pay for
that caisson, and I remember thinking that as I was
working for thirteen dollars per month, and did not
have any surplus, the government would probably
have to take it on what would be called in these days
the instalment plan. But nothing was ever said further about it, the government considering the claim
worthless, I suppose.

March 10, 1862, we made our first march towards
the rebels, and it was a memorable one from the
fact of its being the most disagreeable of any I made
during the war, save one. We broke camp early in
the morning in a cold, drizzling rain, so cold that it
froze as fast as it fell, and moved out near the Centreville pike, where, after waiting for an hour or so,
we finally moved into the road and started towards
Centreville. We made camp that night at Fairfax
Court-House, and next day went as far as Centreville, and found the rebels had left. We remained
here until the 15th, when we were ordered to march,
and back we started towards Washington, turning
off at Bailey's Cross-Roads towards Alexandria, and
finally about seven in the evening drove into a farm-
yard at Cloud's Mill and went into camp, the most
miserable lot of beings you ever saw, nearly frozen,
hungry, and wet through.

My father had sent me a pair of rubber boots to
march in, and I wore them on this march for the first
and last time. I put them on in the morning, tucked
my trousers nicely into them and mounted my horse.
As soon as my clothing became saturated with water

the surplus began to run down into those boots, and my misery commenced. I could not change them, as my shoes were back in my knapsack, and that was inaccessible; but you can imagine that I lost no time after we had unhitched in finding that knapsack and changing my footwear. By the time I had done this the boys had taken the fence in front of the dooryard and built a large fire, and we all hovered around it till morning. At some time during the next forenoon we moved back to our old camp.

April 4th we were ordered to prepare for another march, and we started over the same road. We made our first camp near Fairfax Court-House. On the 5th we reached Manassas, and on the 6th Bristoe Station. That night in camp at Bristoe Station it snowed and was rather cold. I remember I was on guard over horses. Did not have any trouble for the first two hours, but when I was aroused for my second trick, which came about an hour before daylight, I soon became cold and sleepy, and finally went to sleep, in which condition the relief found me. The sergeant entertained me with surmises of what would probably be done to me when it was reported at headquarters. He said that in ninety-nine cases out of one hundred the culprit was shot, and that I could take what comfort I could out of the chance that I might possibly be the one hundreth. I did not feel very happy that morning, and wished I had never become a soldier. I never heard any more from it, and a little later in my experience such small matters did not trouble me much.

We remained here until the 16th, then went to Cattlet Station. Started again on the 18th, and at night reached Falmouth, near Fredericksburg, and went into camp. The next day moved on a little further and went into camp opposite Fredericksburg, on the north bank of the Rappahannock, which proved to be our abiding place until May 25th. On the 26th of May our corps moved across the river and on towards Richmond, our battery remaining in Fredericksburg, camped on a common. The citizens were very bitter, and showed their hatred in various ways, the older ones being careful not to annoy us openly, but through children so small that we could not notice their acts, they sometimes made life almost a burden. I remember passing a fine place that abutted the street, with a high brick wall, on one of the main streets one day, when three or four little fellows, the oldest not more than six, made it very warm for me by throwing gravel at my head as long as I remained within reach.

We left Fredericksburg on the 6th of August and marched to Rappahannock Station, reaching there late in the afternoon of the 8th. To accomplish this we were obliged to keep moving most of the time, halting occasionally for an hour to allow the men to lie down by the roadside and get a little rest.

My recollection of that campaign is, that for long-continued hard marching, and severe fighting, it exceeded anything in my experience. On our arrival at Rappahannock Station we went into position on the north bank of the river. We had heard heavy firing all the afternoon, and knew that a severe battle had been fought. This proved to be the battle of Cedar Mountain, the first in General Pope's campaign.

Some time after dark we were ordered to limber, and pulled out into the road and started on our retreat toward Bull Run. From this time until the

28th we were on the march, moving here, there and
everywhere. What it all meant at that time we did
not know, but have since learned that we were trying
to find Stonewall Jackson, who had in some way become lost,—at least to our generals,—and they could
not find him ; but on the afternoon of the 28th, between five and six o'clock, we found him, or rather
he found us. We were turning into the lots for a
camp, and had some of the horses unharnessed. We
were unaware that the rebels were anywhere near us,
when all at once we were greeted with a tremendous
volley from infantry, which was startling, to say the
least.

We were immediately ordered to hitch up and go
into position, and we had our first opportunity of
showing how we could handle our guns in the pres-
ence of the enemy. For two hours or more we kept
up a very rapid fire, and I think must have done considerable damage. We had the stock of one of our
caissons broken by their fire, and the caisson was
blown up under the direction of Lieutenant Parker.
We lay in this position until about twelve o'clock,
when word was passed that no one was to speak

above a whisper, the drivers mounted, pieces were limbered, and we started silently away.

This battle was called Gainesville from its proximity to a village of that name. We had gone a little way beyond this village towards Groveton, but retraced our steps to the pike that ran to Manassas Junction, to which place we now marched, arriving early on the morning of the 29th. Late in the forenoon we started back, taking the road to Bull Run battle-field. We moved along very slowly, in consequence of the road being occupied by wagon trains. We could hear firing from the fight at Groveton, and were very impatient at our delay How I did fret over it! I was sure that we would be too late, and should have no chance to get at the Johnnies. I was never so anxious afterward; no amount of delay ever disturbed me after my experience of the next day.

About four o'clock we turned off the road, and were ordered forward on the run, and finally went into position on a hill overlooking quite an extent of country toward an unfinished railroad, where Jackson had been fighting our troops since noon. We could see the fighting very plainly from our position, but it was too far away for us to take any part.

I soon began to see the effect of war; wounded men began to pass through our battery, and I became convinced that this was serious business. I remember one poor fellow who passed through our lines three times within two hours on that afternoon, each time with a fresh wound. Twice he had them dressed and went back, but the third time he came back on a stretcher, and we saw no more of him.

We remained all night and until about two o'clock in the afternoon of the next day in our first position. We then moved back across the valley to a position a mile or so off to the rear and left of the old one.

We came into battery by the right flank, which placed my piece, it being the left of the battery, on the right of our line. Two other batteries were placed in position on our right in echelon, a brigade of infantry was brought up, and placed in our rear for support. I remember hearing the captain say to the general commanding the troops in our rear, all he wanted of him was to drive the Johnnies out if they got in between the guns, and to cover his limbering the pieces and taking them off. The general

responded that he should stay there as long as the
battery did.   It is possible that he did stay, but his
men did not, but dusted out long before the battery.

I think that Battery D, as it stood in position
that afternoon, was as fine an organization of the
kind as there was in the service.   By far the larger
part of the men were under twenty-one, active and
thoroughly posted in the drill, and capable of doing
as effective work as it was possible to do.   We had
been drilling nine months, and there were few men
who, if called upon, could not take any position on
the piece and do the work perfectly.

The battery consisted of six brass guns called
" light twelves," or " Napoleons."   They were smooth
bore, and our ammunition was of twelve-pound
shot and shell, spherical in form, and canister, which
consisted of thirty-two quarter-pound iron balls,
contained in a tin case.

Now with such guns and ammunition as these we
had been told, if we would stick to the guns, no troops
could live in front of us as long as our ammunition
lasted ; and on this very afternoon of which I write
we proved the truth of that assertion.   We had been

in position perhaps a half hour, when we had orders
to begin shelling the woods that were in our front.
We had seen great clouds of dust in the direction of
Manassas Junction pike, and very soon it was said
that Longstreet had come up on Jackson's right, and
was swinging in on our left with the intent of doub-
ling us up.

Soon they began to show themselves, and we kept
up a pretty lively fire on them with solid shot and
shell.   We must have made it rather warm for them,
as the writer noticed they covered as soon as
we got the range.   They suddenly appeared in our
front and formed for a charge, two or three regiments
front, and several lines in the rear at close intervals.
Soon the gunners called for canister, and we began
to send that into them; then we double-shotted it,
breaking off the cartridge from one of the cases and
ramming it home on top of the other.   Our gunners
had been taught that in firing canister to prevent
wasting it, it was best in close action to ricochet it,
having it strike the ground just far enough in front
of the enemy to have its rebound reach them breast
high.   Now imagine the execution of six guns,
handled by cool gunners as these were, and can you

wonder that they fell back? They soon rallied,
however, and came for us again, and this time we
staid with them until we used every round of ammunition we had, then limbered and started for the
rear, taking off every piece and caisson. The batteries on our right had all been captured or driven off,
and when we started for the rear I do not remember
seeing any of our troops anywhere near us.

I went into this action as driver of the lead team
on the caisson, but early in the fight one of the lead
horses of the piece was disabled, and my team was
taken to replace it. This left me without anything
to do for awhile, but as the limber of the piece was
soon emptied, I found plenty of employment in
bringing up ammunition from the caisson to the
piece. I finished this just about the time that the
Johnnies started on the second and last charge, and
I had nothing to do but stand around and watch
things, which was just the thing I did not care to do.
The rebels had planted some artillery on our left,
and opened a heavy fire on us with shell and canister, and it seemed to me that the air was full of
screeching shells ; then canister would come bound-

3

ing through our battery, so nearly spent that I could
watch them, and it did seem to me that if I got out
of that place alive it would be a miracle. Just then
I heard the order, "Limber to the rear," and I remember calling as loudly as I could, repeating the
order, to Corporal George Eldred.

The caisson had left while I was calling to Eldred,
and the piece followed closely. Corporal Eldred had
passed me on his way to the rear, he being the fast-
est runner, but we had not gone a hundred yards
when I heard a shell coming behind us, and just as
it reached a point directly over my head and not a
great way above it, it burst, the concussion from it
nearly knocking me down. Then I heard the whirr
of a part of the shell as it flew in front of me, and
the thud as it struck poor Eldred square in the back
of his neck. I shall never forget the sound that he
made as he fell forward, nor the last sight I had of
him as I passed a moment later. His head was bent
under him a little, showing a great gaping wound in
his neck.

A singular thing in connection with his death was
the fact that he had always declared he should be

killed in the first battle that he should participate in.
The boys had tried to laugh him out of the notion.
The night before, in our old position, some half dozen
of us were lying around the gun, and Eldred had
again assured us that if we went into action on the
morrow he would be dead at night, and no amount
of chaffing could dispel his melancholy.

Soon I caught the piece and jumped on the trail,
which I had hardly done when we jumped a wide ditch,
and I thought I had been hit with a hundred pound
shell, but clung to my place, and rode along until
the battery was halted and ordered into position
again, for what purpose I never could understand,
as we had no ammunition, unless as a piece of unadul-
terated bluff.   Nothing came of it, however, as
the Johnnies seemed to have had enough for the pres-
ent, and did not follow us far.   We limbered again,
and started towards the stone bridge, which, fortunately, we succeeded in crossing without much
trouble, and moved on a mile or so towards Centreville.   We then turned into a lot and took a short
rest.   Later we moved on to Centreville, where we
unhitched and fed our horses, made some coffee for

ourselves, and I lay down to get a little sleep. It
was rather disturbed, however, as I was continually
hearing shot and shell whirling around me, and I
frequently awoke trying to dodge them. We remained all day at Centreville replenishing our ammunition, and the next day,—September 1st,—started
for Washington. When just about half way between Centreville and Fairfax Court House, Jackson
opened upon our corps at a place called Chantilly,
and a very severe battle, but of short duration, took
place. The battery did not take part in this, as the
fighting was done principally by infantry. How it
rained that night; we were wet through, but were
so tired that we spread our blankets and lay down
on the wet ground, threw the blankets over our
heads, and were soon fast asleep.

The next day we moved on, and at night reached
the vicinity of our old camp at Munson's Hill, very
tired, very hungry, and very much discouraged. A
brigade of infantry made camp on the opposite side
of the road from us, and I remember hearing some
of them say that it was fortunate that the officers
had concluded to make camp just as they did, as

they could not have gone a step farther for any
one.

Just about dark on this same night—September
2d,—we heard cheering away off down the road towards Alexandria. Of course we were all very anxious to know its meaning. Soon we could distin-
guish a large body of horsemen approaching, and the
troops on either side would rush out into the road
and cheer with all their might. We rushed with the
rest, and when the cavalcade got near enough we
saw that it was General McClellan and staff, and word
was passed that General Pope had been relieved and
McClellan had assumed command again. What a
transformation took place in those troops ! All signs
of discouragement had passed away, and I fully be-
lieve, if he had asked them, tired as they were, to
recommence their march, or go into fight that night,
they would willingly have obeyed, such was their
confidence in him.

For a week or more we remained around the de-
fences of Washington, doing a little picket duty now
and then. Finally, on the 6th of September, we left
camp near Dupont about 9 p. m., and marched to

Washington, passing through the city about midnight, and on the 7th we camped about twelve miles beyond the city, on the Maryland side of the Potomac river. We remained here until the 10th, and then went to Lisbon. We reached New Market on the 12th, and went to Frederick City the next day. Here we began to skirmish with the rebels, and on the 14th and 15th our troops had a severe fight with them at South Mountain. We did not become engaged, but the fighting was in plain view from our position. We had a little excitement on the afternoon of the 14th. Our men, together with some from a regiment that lay just across the road from us, participated in a raid upon a sutler who was unfortunate enough to happen along just as they were very much in need of something nice to eat, but had no money to pay for it, and his possessions were speedily reduced to the horse and running-gear of his wagon, without any collateral to show for it. He complained at headquarters, and every effort was made to find his goods, but not a single thing was found.

The battery moved down towards Sharpsburg on the 16th, and took a position soon after dark. The

rebels shelled us until about 9.30 P. M., and it made
a very pretty display as the shells passed through
the air, leaving a track of fire behind, and I think we
should have enjoyed it if it had not been so dangerous.
I laid down that night on the top of a caisson, about
ten o'clock, and went to sleep. Just about daylight
I awoke with a start. I think that the whizz of a
shot must have awoke me ; at any rate, just as I
raised my head, one passed over me, so close that I
thought at the time it could not have been more than
an inch above me ; but I suppose it really was several feet. I jumped down from that box quickly, and
for ten or fifteen minutes the Johnnies threw a stream
of shells through our battery. They had a perfect
range on our position, and for a little while made it
very warm, but we happened to have in position on
that hill about twenty guns, unlimbered and ready for
action, and it took but two or three minutes for our
cannoneers to get to them, and then in a moment
twenty projectiles of various kinds and size were fly-
ing towards that rebel battery. That treatment was
kept up until we had the pleasure of seeing them
limber and run away.

We remained in our position, if I remember rightly, until shortly after noon, firing whenever we could see anything to fire at, and watching the fight on our front. Immediately in front was a thin belt of woods, and just beyond this an extensive corn-field, in which was done as stubborn fighting as I ever saw. First one side, then the other, would hold possession of it, charging back and forth, leaving the dead and wounded on the field until they lay in windrows from one side of the corn-field to the other. Early in the forenoon, if my memory serves me, Captain Monroe was ordered to take his pieces, leaving the caissons, and go down through that corn-field to the farther end, and if possible, silence a rebel battery that was giving our men trouble. Well, we started, and what an awful journey it was. We no sooner reached the field than we were greeted with the groans of the wounded, and some of them that lay in our way had to be moved to one side. Some were horribly mangled. Such sights as these, and the constant zip of the Minies (which sound I always disliked very much more than that of the shot and shell) had completely unmanned me, so that when we had unlimbered and I was called upon to cut a fuse, I found that my right hand was trembling so I could not use the cutter, and I called upon the wheel driver to help me. In a moment it had passed away and I was myself again. Our gun did the most rapid firing here that I ever knew it to do, and in a very short time we had silenced those guns. I learned afterwards that we knocked that battery almost to pieces. We then limbered and went a little way back and halted, while one of our pieces was prolonged off. The sharpshooters had nearly used up both men and horses on that gun, in attempting to limber. I think they shot five men, one after the other, just as fast as they attempted to take hold of the trail, but the men succeeded in attaching the prolong to it and dragging it off by hand.

While we lay in that hollow, a division of nine months troops came out into the field. We knew they were fresh from home by the newness of their uniforms and the fresh look about their colors, and also from their full ranks. They marched out in fine style, stepping over the old line of battle where the men were lying down, and charged towards a stone

wall behind which the rebels were. Some of the old
fellows chaffed them a little by asking where they
were going, and telling them to be sure and not go
beyond the stone wall; and they did not, but very
near it, and a brave fight they made of it, so brave
that the vets gave them a rousing cheer when at last
they gave way and fell back.

Soon after this we went back to our old position.
It was getting dark by this time, and pretty soon the
firing ceased and everything was quiet except a shot
now and then from the pickets. Thus ended the
battle of Sharpsburg, or Antietam.

From our standpoint we could not tell whether it
had been a victory for us or not. We certainly did
not suppose it was over, and expected to commence
fighting at daylight; but next morning, everything
being quiet, we investigated and found that there
had been a cessation of hostilities asked for and
granted, for the purpose of attending to the wounded.
All that day we remained in position, expecting
every moment to be called upon. It began to be ru-
mored around that Lee was getting away as fast as
he could, and some very forcible remarks were made

about allowing him to get away without making any
attempt to crush his army. Every soldier that I
heard express himself was in favor of fighting.
General McClellan has said that the troops were not
in condition to follow, but that was certainly not the
case with any of the troops around us.

On the 19th we started, as we supposed, in pur-
suit, but only marched a short distance and went into
camp. After we had finished our camp duties, some
of us went back over the battle-field, and I shall
never forget what I saw there. A great many of the
dead lay just as they had fallen two days before.
Burial parties were engaged in digging trenches fifteen to twenty feet long, about six or eight feet in
width, and four to five feet deep, in which they
would lay the dead as closely as possible, then cover
them up. Most of the bodies were in a terrible
state. It had rained the previous night, and the sun
coming out very hot the next morning hastened mor-
tification, turning the exposed parts of the bodies
black, while they were swollen to two or three
times the natural size. I remember seeing a young
boy, who had evidently been mortally wounded and

had dragged himself up near a stone wall to die. He had taken a daguerreotype of his mother from his pocket for a last look before he died. A horse sitting up like a dog, with his nose deeply imbedded in a hay-stack, dead, was among the singular things I remember to have seen on that field.

September 20th we moved near Sharpsburg and made camp, in which we remained until October 20th. During this time we were reviewed by President Lincoln.

October 20th we went to Brownville, where we remained three days, camping with the division of artillery, and we expected to build winter quarters here, but on the 23d were ordered to pack up, and after marching until 9 P. M., went into camp in a mud-hole. On the 27th moved three or four miles to Crompton's Pass, and on the 28th continued through the Pass and camped near Knoxville, Md. On the 30th went to Berlin. November 1st went into Virginia and camped at Berryville, and on the road passed the Seventh Rhode Island. On the 3d camped at Bloomfield, meeting the Fourth Rhode Island. On the 5th marched to Rectortown, thence to Warrentown. On this day we went into position in a furious snow storm. On the 10th McClellan's farewell address was read to us on parade, the army was reviewed, and Burnside took command and his address was also read. On the 11th we marched to Waterloo, remained here until the 17th, when we went to Morristown, where it was said we were to quarter for the winter; but on the 22d we marched to Brook Station on the Fredericksburg and Aquia Creek Railroad, and from there back to Waterloo. From this time until December 4th we remained at Waterloo. The weather was very cold and stormy. On the 4th we were ordered to pack up, but it began snowing very fast, and the order was countermanded. It stormed all day and part of the next, the snow was three to four inches deep. We did not move until the 7th, when we went four or five miles, and the battery became so mired we were obliged to make camp where we were. Rations were very short at this time, and we could not get a square meal. On the 8th we went to Fredericksburg, and on the 9th made camp opposite the city. On the 11th the battle of Freder-

4

icksburg was opened. Some time before daylight
on the morning of the 11th, artillery had been posted
along the heights opposite the city, reaching from
one end to the other of the town. I do not know
just how many guns were in that line, but should say
from seventy-five to one hundred. About 5 A. M. a
signal gun was fired, and then they all opened, and
for three hours there was a constant roar from these
guns that fairly shook the earth.

There was a brigade of rebel sharpshooters in
Fredericksburg at this time, and for a long time they
effectually prevented the laying of our pontoon
bridge. Our men would be shot down as soon as
they showed themselves, and finally orders were
given for the artillery to concentrate their fire upon
them, and at the same time volunteers were called
for to cross in the boats and drive them, which was
speedily done, when the bridge was laid and our
army began to cross, and heavy skirmishing was car-
ried on all day. The next day a very heavy force
was thrown across the river, and our battery went
with it. We were not called upon, however, but
lay all that day and until just before dark the next,

under cover of the buildings in the streets of Fredericksburg.   Just before dark on the 14th we were
ordered into position on a hill just beyond the town,
and opened fire; but no sooner had we done so than
a perfect shower of shells and Minies, we being
within rifle range of their works, poured in upon us,
which shut us up about as quick as you could a
jacknife, and we were very glad to seek the protection of the bank, coming back, however, as soon as
it ceased, and opened fire again, with the result of
reproducing the shower.   About this time we discovered a new danger.   It seems that just before we
occupied this position another battery had moved
out of it, and they had piled their ammunition up
near their pieces to have it handy.   It was so dark
when we drove in that we did not see it.   We had
been firing over it, and soon a spark from our guns
set fire to a fuse, and we suddenly found ourselves
in a hornet's nest.   Some of us did not have time
to get under the bank, and for fifteen minutes or so
we hugged the ground for dear life.   After remaining here long enough to become satisfied that the
only service we could do would be to act as a target

for the Johnnies to draw their fire, we were ordered
out of it, and went down into the town, seeking
shelter under the lee of the buildings. We remained
here until about 2 A. M. of the 15th, when we re-
crossed the river and returned to our old camp. This
ended the battle of Fredericksburg, lasting four
days, every one of which was filled with hard fighting. The battery was under fire constantly, as the
rebels kept up an almost constant fire, and so accu-
rate that we could but remark the wonderful im-
provement their gunners had made. Later we were
able to understand the reason of this remarkable
shooting; diagrams of the surrounding country were
found, the prominent points in which, such as streets
in the city, farm-houses and intersecting of roads,
etc., had been marked with the degrees of elevation
necessary to reach the spot, the result of an actual
survey. A perfect dub could have made a good shot
with such help.

We moved on the 17th of December back about a
mile and a half, into a grove, and began to build our
winter quarters. My chum, Peter Botter, and myself, finished ours in a day or two, and made our-

selves quite comfortable. Our house consisted of a cellar about a foot and a half deep, six feet long and four feet wide, boxed around with pine slabs. Then the dirt was tamped hard around the outside of the slabs, a ridge-pole raised in the crotch of two upright poles and covered with our shelter tent ; a mud chimney was built on the outside, the tent being tacked tightly around the fire-place. We had a bunk on either side, raised from the ground and filled with boughs. When the house was done and we had built a good rousing fire in the fire-place, we were just as comfortable and happy as it was possible for soldiers to be.

From this time until the 6th of February, 1863, we spent our time in performing the ordinary duties of the soldier, such as drill, having inspections, etc., varied between January 10th and 21st by being under marching orders for the purpose of crossing the Rappahannock river on an expedition against the rebels, and a large part of the army did leave their quarters and make the attempt, but the weather was so bad and the roads so muddy the idea was abandoned, and the troops returned to their old camps, complet-

ing what is known in history as the celebrated mud
march. Our battery was fortunate enough not to have
left their camp.

On the 6th of February we received orders to pack
up and be ready to march in an hour, and at 8 A. M.
we pulled out of our winter camp and started for
Belle Plain, on the Potomac river, about twelve
miles distant. It rained very fast and the roads
were exceedingly muddy, so that the very best we
could do in all day was between five and six miles.
The pieces and caissons would become fast in the
mud, and we would have to double our teams to pull
them out. You may imagine our condition : tired
out, wet through, and no way of protecting ourselves from the cold storm, which continued through
the night. We succeeded after great difficulty in
pulling the guns and caissons through to the landing
on the next day, but the battery wagons and forge
not having arrived, six teams of horses were sent
back after them, and they were found about five
miles back, the forge being bottom side up in a
creek, having run off the bridge the night before.
After four or five hours of hard work, we got it out,

and after great trials and tribulations we finally
landed the battery at our destination, having been
just three days going twelve miles.

At two o'clock A. M., on the 9th, we commenced
to load the battery on canal boats; by 9 A. M. were
loaded, and at 4 P. M. started down the river in tow
of a steamer, but went only a little way because of
the rough condition of the river. On the 11th we
again started, but as we reached the bay the captain
decided that it would not be wise to attempt to
cross in the canal boats, so we made harbor at St.
Mary's, where we lay until the 13th. The oysters
were so plenty here that the boys would take a boat
and row over to the rocks, returning in a very short
time with several bushels. We did just have a feast
while here; we ate them in every style, raw, fried
and stewed. I can remember even after this lapse
of time how good they tasted.

At daylight on the 13th we started once more, but
after running down opposite Point Lookout were
obliged to lay to again on account of the weather,
until three o'clock, the wind having gone down, we
pushed on and reached Hampton Roads at daylight

on the morning of the 14th, and immediately disembarked, and the next day went into camp near Hampton, which must have been a beautiful place in its day, before the rebel General Magruder burned it. When we arrived it was largely occupied by negroes, and here it was that I first attended a genuine negro meeting. The boys used to go frequently, more as a place of amusement, I am afraid, than for a better purpose.

From the 15th of February until the 4th of March, we remained in camp at Hampton, with very little to vary the monotony.

On the 14th of March we moved over to Newport News, where we remained until the 19th, and at six o'clock on that day we started for Fortress Munroe, when a furious snow storm came on and we made camp at Hampton. How it did snow that day! At night it was eight to ten inches deep, but next day we pushed on to the Fort. From the wharf at this point we embarked on a schooner, and early on the 22d sailed for Baltimore, arriving there at sunrise on the morning of the 23d, and commenced immediately to load the battery on cars. At three P. M.

we left Baltimore over the Baltimore and Ohio
railroad. It was my duty to look after the pieces
and caissons on the first two flat cars until midnight,
at which time I was relieved. We had just changed
engines for the middle section run and I had made
friends with the engineer, who invited me to take a
seat in the cab. It was a camel-back engine, and
the cab was on the top of the boiler. How warm
and nice it did seem to me. When we arrived at
the middle of his run, which was on the very tip-top
of the Cumberland Mountains, he invited me to go
to lunch with him, which I gladly accepted. I remember that I thought it a great lay-out, and just
filled myself. As we were leaving to return to the
train he handed me a dozen sandwiches, and I tried
to convince him that I thought he had the biggest
heart of any man I ever met. Whether I succeeded
or not I do not know. We ran all the night of the
24th and all day the 25th, arriving at Parkersburg
on the 26th, when we loaded our battery on a
steamboat, and on the 27th started down the Ohio
river. We had a delightful sail down the river, and
ran our nose on the bank some six or eight miles

above Cincinnati about eight o'clock P. M. on the
28th. The next day we ran down to Covington and
transferred the battery from the boat to cars. Un-
fortunately for the physical condition of many of
the men in our company, there lay around that depot
several hundred barrels of whiskey, brand-new stuff
just from the distillery, almost white in color, and
containing enough fusel oil in a wine-glass to kill a
man, unless he was of the toughest kind. Well,
there were about twenty-five men in the company,
who the moment they discovered that whiskey were
determined to possess as much of it as possible; so
at about ten P. M., when we rolled out of that depot,
we had a barrel of the whiskey aboard, and just as
soon as we were fairly under way, the head was
knocked in, and very soon every canteen, water-
bucket, in fact everything that would hold whiskey
was full of the stuff. The sequel of this bounteous
supply, rivers of whiskey as it were, was, that on
our arrival at Lexington next morning, about eight
o'clock, nearly fifty per cent. of our men were incapacitated for duty, and it consequently fell upon the
temperance element of the company to do all the work.

The battery was finally unloaded, and we moved about a half mile from the city and went into camp. Two of our sergeants, Taft and Sullivan, were reduced to the ranks for misdemeanor on the trip. We remained in this camp from March 30th to April 8th, doing general duties, visiting the city, and among other places the plantation formerly belonging to Henry Clay. Late in the evening of the 8th we were ordered to pack up, and marched eight or ten miles, bivouacked for the night, and starting early next morning, made twenty miles, reaching Camp Dick Robinson, where we remained until the 26th.

My recollections of this camp are very pleasant. Just below and across the pike from our camp stood the Hoskins mansion, a spacious house built very much after the style of the houses on the large and wealthy plantations of the South; very large on the ground, and having a wide piazza on all sides. At the time of our visit the house was occupied by Mrs. Hoskins and some half-dozen slaves, all that remained, as she told me, of seventy-five that she owned at the beginning of the war. I made an informal call on the lady very soon after our arrival. She

received me pleasantly, and as dinner was ready she
invited me to sit up. Our conversation during this
meal developed the fact that madam had about everything that was good to eat on her plantation, but no
money. We had money, but nothing good to eat, and
it was arranged that some dozen or more of us should
part with a dollar each day and receive two good
meals in exchange. Mrs. Hoskins fulfilled her con-
tract to the letter, and rounded it off by spreading
for us on the last day of our stay a perfect banquet.

We left Camp Dick Robinson on the 26th of April,
and from then until May 7th spent our time in
marching around the country, visiting Stamford,
Columbia and Carpenter Creek, for the most part
over fine roads and through a beautiful country.

We reached Somerset on the evening of the 7th,
and remained there until June 3d, leaving there at
sunrise, and marched towards Lexington, stopping
over one day at Stamford to receive pay, and reached
Lexington about ten A. M. on the 8th. Began immediately to load the battery on the cars to commence
our journey to Vicksburg, but after having nearly
loaded the battery the order was countermanded, and

we went about three miles from Lexington and
camped. That night Louis La Fount, a member of our
company, was brought to camp, dead, having fallen
or been thrown down stairs in the guard-house at
Lexington, and his neck broken. On the 10th we
marched to Nicholsville, and the next day went
about five miles to Camp Nelson and remained there
until July 12th, a most delightful camp in a very
beautiful country. On the 4th we celebrated by firing a salute, and in order to make as much noise as
we could we cut grass, and putting as much as we
dared in the guns, rammed it home, and instead of
being satisfied with the noise of one gun at a time,
we fired by battery : that is, the six guns in unison,
and as you may imagine the report was a loud one.
In the afternoon we went to the village and celebrated. On the 5th rumors of Morgan's approach
began to fly about, and soon the citizens began to
drive in their horses and cattle. For our protection
we placed four pieces in position, and the infantry
threw up earth works ; but Morgan was not the fellow to come when he knew we were prepared to
receive him. He gave us a wide berth, and by the
5

11th we knew that he had crossed the Ohio river into Ohio. On the 12th we started at 9 A. M. for Lexington, and loaded the battery with all possible haste, starting as soon as loaded for Cincinnati, arriving at Covington at 8 A. M., on the 13th, and immediately crossed the river into Cincinnati. The city was very much excited, as Morgan was reported to be within ten or twelve miles, and the citizens expected to see him ride into their streets at any moment.

Our battery was the only veteran organization in the town at this time, and from the moment we landed on that levee we began to receive an ovation. The citizens met us with open arms, as it were, and seemed to feel as though there was some chance of taking care of Morgan now that we had come. The demonstration grew as we got farther into the city, and when we crossed the Rhine, a canal that run through the town, it reached a climax. This part of the city was largely occupied by Germans. There was a lager beer saloon on every corner, and sometimes one or two between. As soon as we reached their neighborhood the saloon-keepers came out to us with both hands filled with glasses of beer. Most of us

indulged once, some two or three times, and others so
frequently that when we arrived in camp on the out-
skirts of the city, we found that history had repeated
itself, and the temperance men had to do the duty.
The next morning the battery was divided into sec-
tions and sent out on the three prominent roads
entering the city from the north.   My section went
about two or three miles, and came into position on
a pike.   We were supported by the Washington
Rifles, good fellows—some of them.   Their relatives
and friends in the city sent them a large load of nice
things, which they shared with us.   We had rare
fun the next morning, before light, stopping the
market wagons as they came along on their way to
market.   They were considerably frightened at seeing two cannon in position in the road, and when
they were halted by the sentry, the sergeant of the
guard called, and the demand made that their wagons should be searched for contraband goods, they
were too amazed to resist, and we would go through
the wagons, taking a few bunches of grapes for our
trouble.   On the 17th we moved back into the city.
On the 19th we hitched up and marched through the

city, visiting the various market-places as a sort of
intimidating act, there being some indication of a
riot.   The next day, thanks to General Burnside,
we were ordered to Ninth Street, and made a novel
camp.   The pieces and caissons were placed in a
wagon-yard, the horses in a stable, and the men
quartered in a hall, and from this time until the 10th
of August we enjoyed ourselves very much.   We
had little to do, Sunday morning inspection on the
levee being about the only real duty that was re-
quired of us.

The citizens on Ninth Street invited us out to tea,
detachment at a time, and entertained us in fine style.
All this was immensely enjoyed by us, but we knew
that it was not soldiering, in the full acceptance of
the term, and that it must end, which it did on the
10th of August.   We crossed the river to Covington, boarded the cars and were taken to Lexington,
from which place we were to commence our march
of 250 miles over the Cumberland mountains into
East Tennessee, where we passed through a winter's
campaign of suffering and privations that was not
experienced by any other army during the war.

www.ingramcontent.com/pod-product-compliance
Lightning Source LLC
Chambersburg PA
CBHW022015190326
41519CB00010B/1532